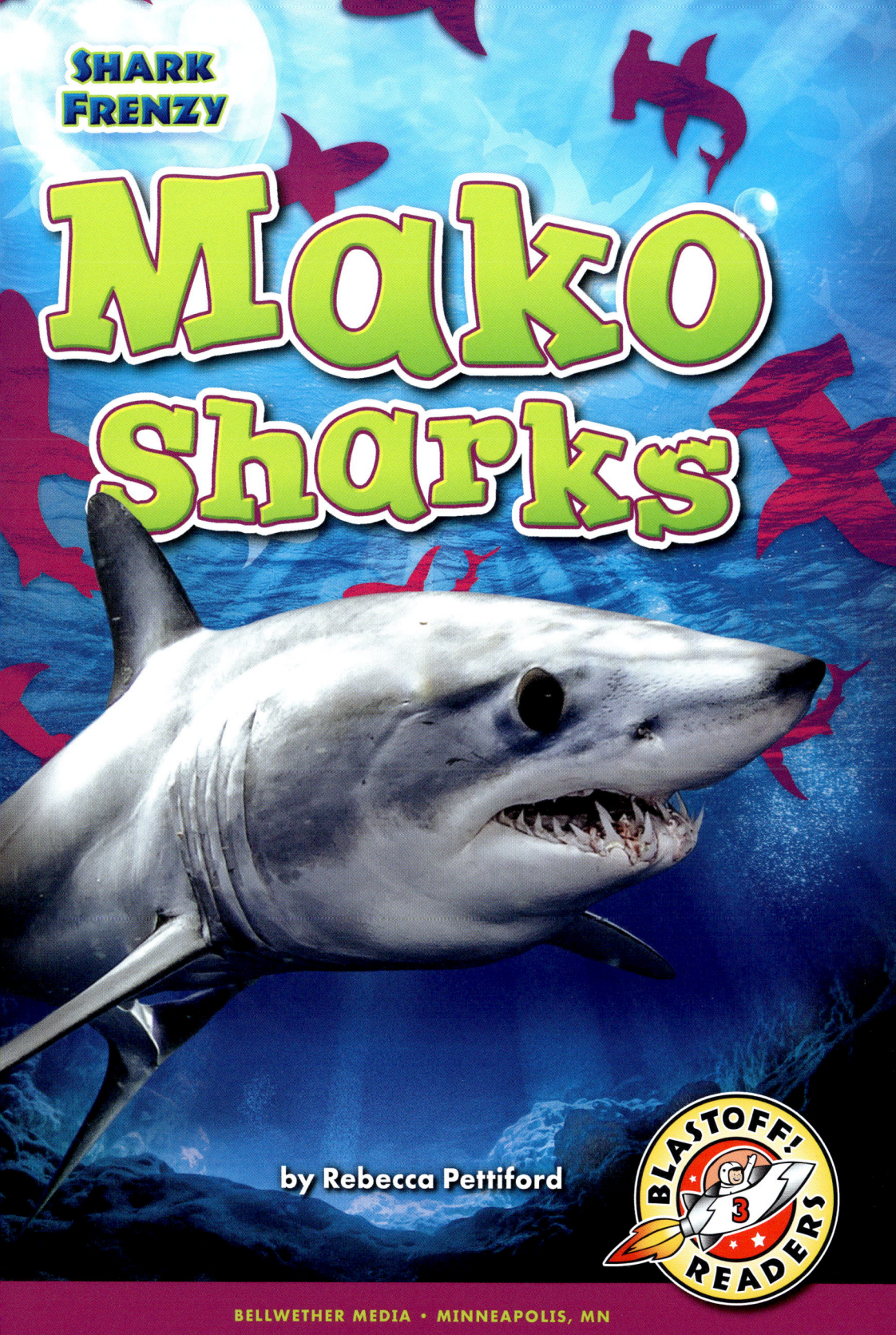

SHARK FRENZY
Mako Sharks

by Rebecca Pettiford

BLASTOFF! READERS 3

BELLWETHER MEDIA • MINNEAPOLIS, MN

Blastoff! Readers are carefully developed by literacy experts to build reading stamina and move students toward fluency by combining standards-based content with developmentally appropriate text.

LEVELS

 **Level 1** provides the most support through repetition of high-frequency words, light text, predictable sentence patterns, and strong visual support.

 **Level 2** offers early readers a bit more challenge through varied sentences, increased text load, and text-supportive special features.

 **Level 3** advances early-fluent readers toward fluency through increased text load, less reliance on photos, advancing concepts, longer sentences, and more complex special features.

★ **Blastoff! Universe**

Reading Level

 Grade K   Grades 1–3  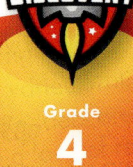 Grade 4

This edition first published in 2021 by Bellwether Media, Inc.

No part of this publication may be reproduced in whole or in part without written permission of the publisher. For information regarding permission, write to Bellwether Media, Inc., Attention: Permissions Department, 6012 Blue Circle Drive, Minnetonka, MN 55343.

Library of Congress Cataloging-in-Publication Data

Names: Pettiford, Rebecca, author.
Title: Mako sharks / Rebecca Pettiford.
Description: Minneapolis, MN : Bellwether Media, 2021. | Series: Blastoff! Readers: Shark frenzy | Includes bibliographical references and index. | Audience: Ages 5-8. | Audience: Grades 2-3. | Summary: "Simple text and full-color photography introduce beginning readers to mako sharks. Developed by literacy experts for students in kindergarten through third grade"– Provided by publisher.
Identifiers: LCCN 2020001637 (print) | LCCN 2020001638 (ebook) | ISBN 9781644872475 (library binding) | ISBN 9781681037103 (ebook)
Subjects: LCSH: Mako sharks–Juvenile literature. | Sharks–Juvenile literature.
Classification: LCC QL638.95.L3 P483 2021 (print) | LCC QL638.95.L3 (ebook) | DDC 597.3/3-dc23
LC record available at https://lccn.loc.gov/2020001637
LC ebook record available at https://lccn.loc.gov/2020001638

Text copyright © 2021 by Bellwether Media, Inc. BLASTOFF! READERS and associated logos are trademarks and/or registered trademarks of Bellwether Media, Inc.

Editor: Rebecca Sabelko   Designer: Kathleen Petelinsek

Printed in the United States of America, North Mankato, MN.

Table of Contents

What Are Mako Sharks?	4
Speed Racers	8
Fast Hunters	14
Deep Dive on the Shortfin Mako Shark	20
Glossary	22
To Learn More	23
Index	24

What Are Mako Sharks?

shortfin mako shark

Mako sharks are the fastest sharks in the world. They can reach speeds up to 60 miles (97 kilometers) per hour!

These **carnivores** live in oceans all over the world.

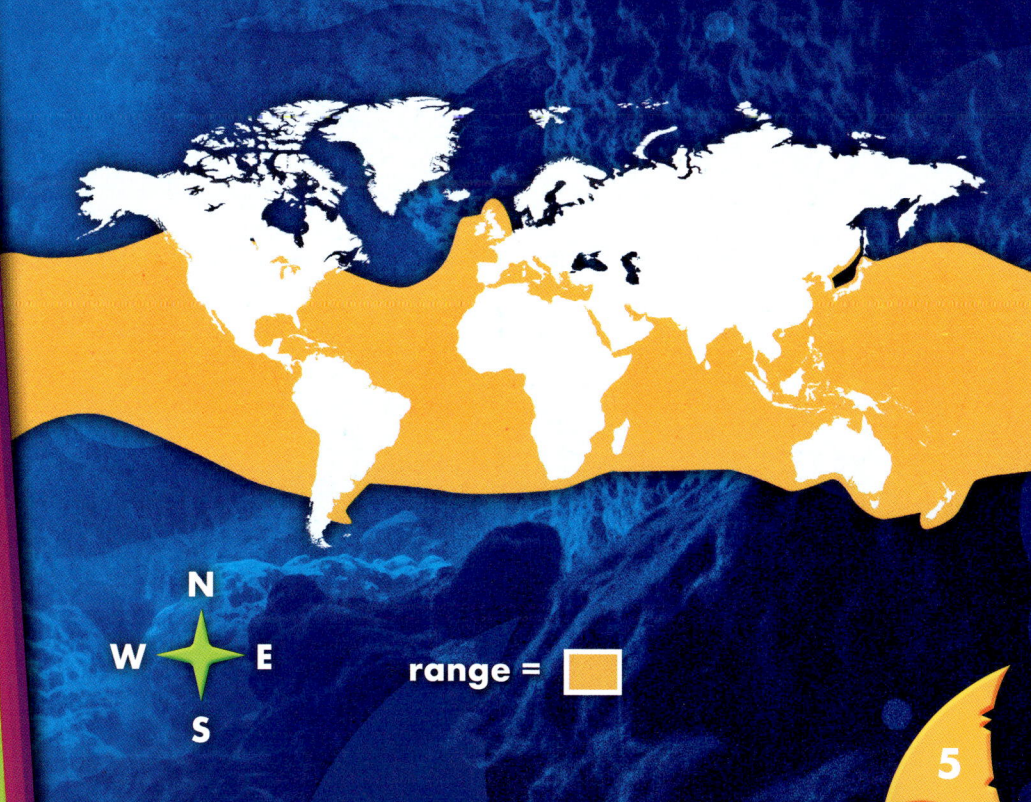

Shortfin Mako Shark Range

range =

Shortfin makos are the most common mako **species**. They have short **pectoral fins** and small eyes. Longfin makos have longer fins and bigger eyes.

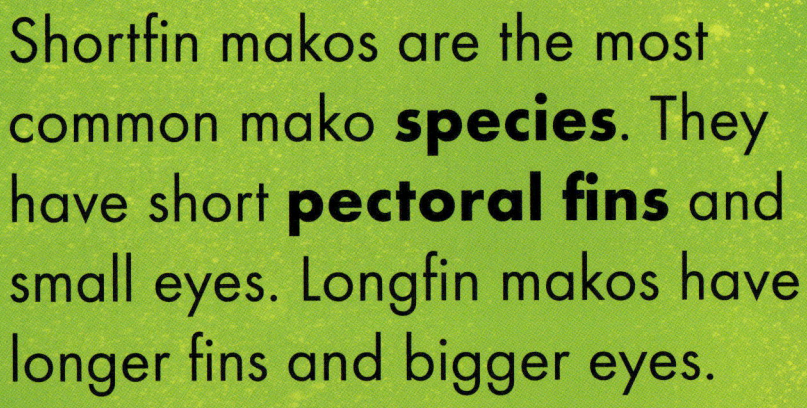

pectoral fin

longfin mako shark

Both species are **endangered**. Overfishing is reducing their numbers.

Speed Racers

Mako sharks are in a group called **mackerel sharks**. Great white sharks are in this group, too.

Like all mackerel sharks, mako sharks are built for speed. Their **torpedo** shape and powerful tails drive the sharks through the water.

Identify a Mako Shark

- torpedo-shaped body
- pointed snout
- pectoral fin

Mako sharks can reach up to 14 feet (4.3 meters) long. They have dark blue backs and white bellies. Five large **gills** help them breathe.

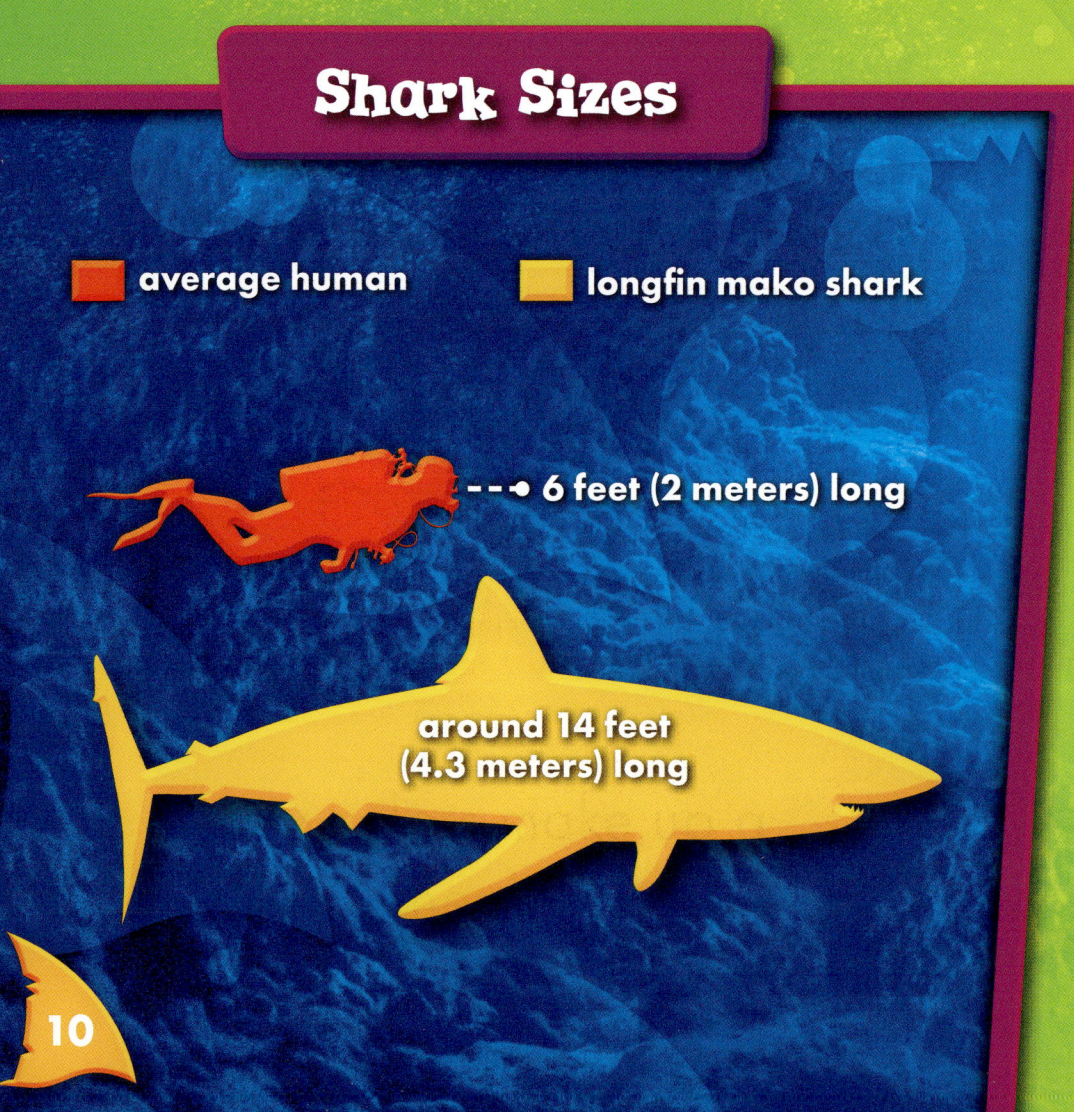

Shark Sizes

- average human
- longfin mako shark

6 feet (2 meters) long

around 14 feet (4.3 meters) long

Mako sharks have pointed **snouts**. Their long, sharp teeth stick out of their mouths. The teeth look like knives!

Parts of shortfin makos' bodies are warm-blooded. This allows the sharks to swim fast in cold waters.

Staying warm requires **energy**. To make energy, makos have to eat a lot. They will swim far to find food.

Fast Hunters

Mako sharks need to use speed and **intelligence** to hunt. Some of their **prey** can swim almost 70 miles (113 kilometers) per hour.

The sharks look for fish and squids. They also hunt dolphins and seabirds.

Mako Shark Diet

fish

dolphins

squids

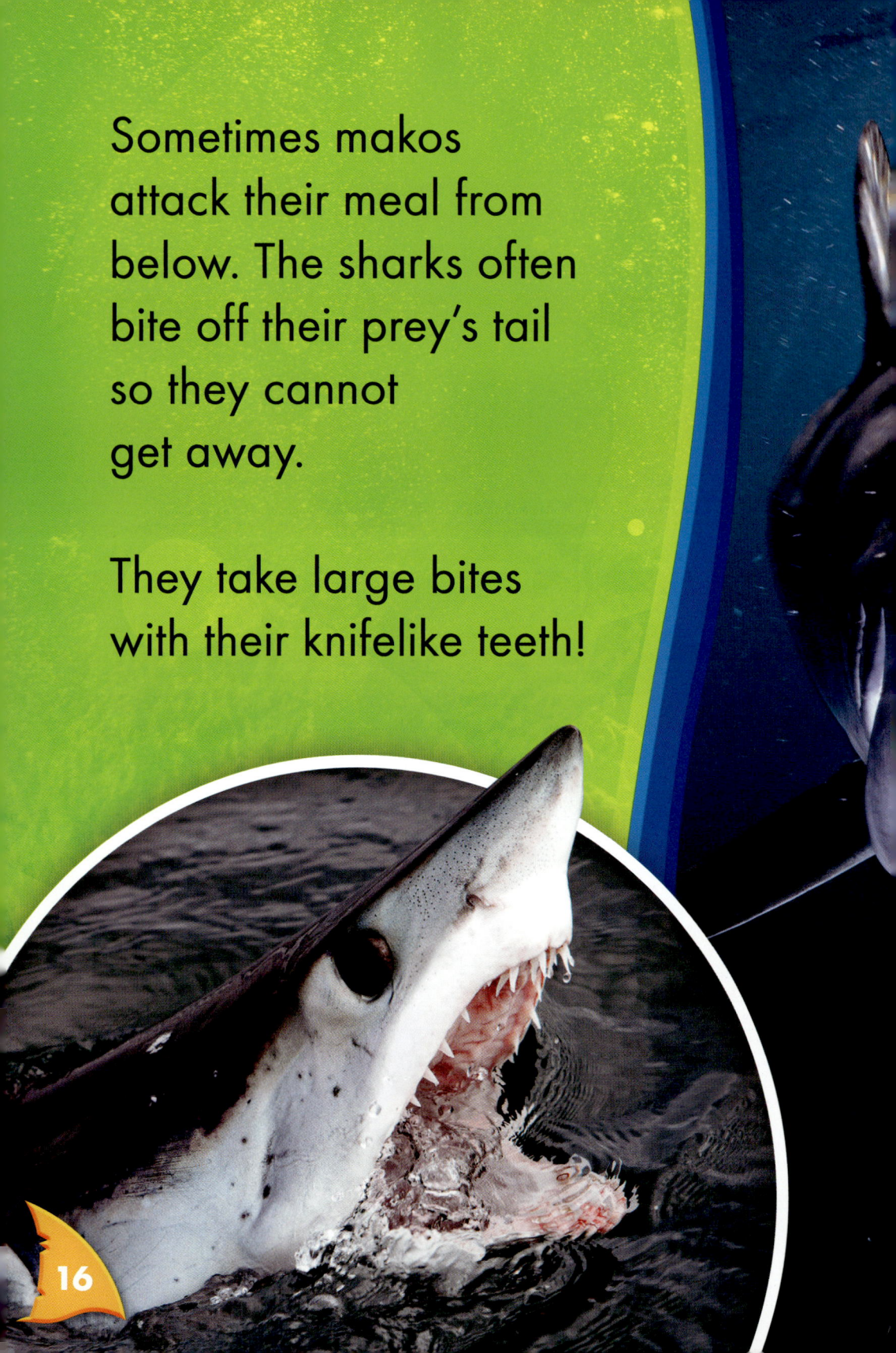

Sometimes makos attack their meal from below. The sharks often bite off their prey's tail so they cannot get away.

They take large bites with their knifelike teeth!

Young mako sharks can fall prey to larger sharks and **orcas**. But adult makos are **apex predators**.

Not many animals can outswim these ocean carnivores!

Deep Dive on the Shortfin Mako Shark

LIFE SPAN:
up to **32 years**

LENGTH:
up to **13 feet (4 meters)** long

WEIGHT:
around **330 pounds (150 kilograms)**

TOP SPEED:
up to **60 miles (97 kilometers)** per hour

DEPTH RANGE:
up to **1,640 feet (500 meters)**

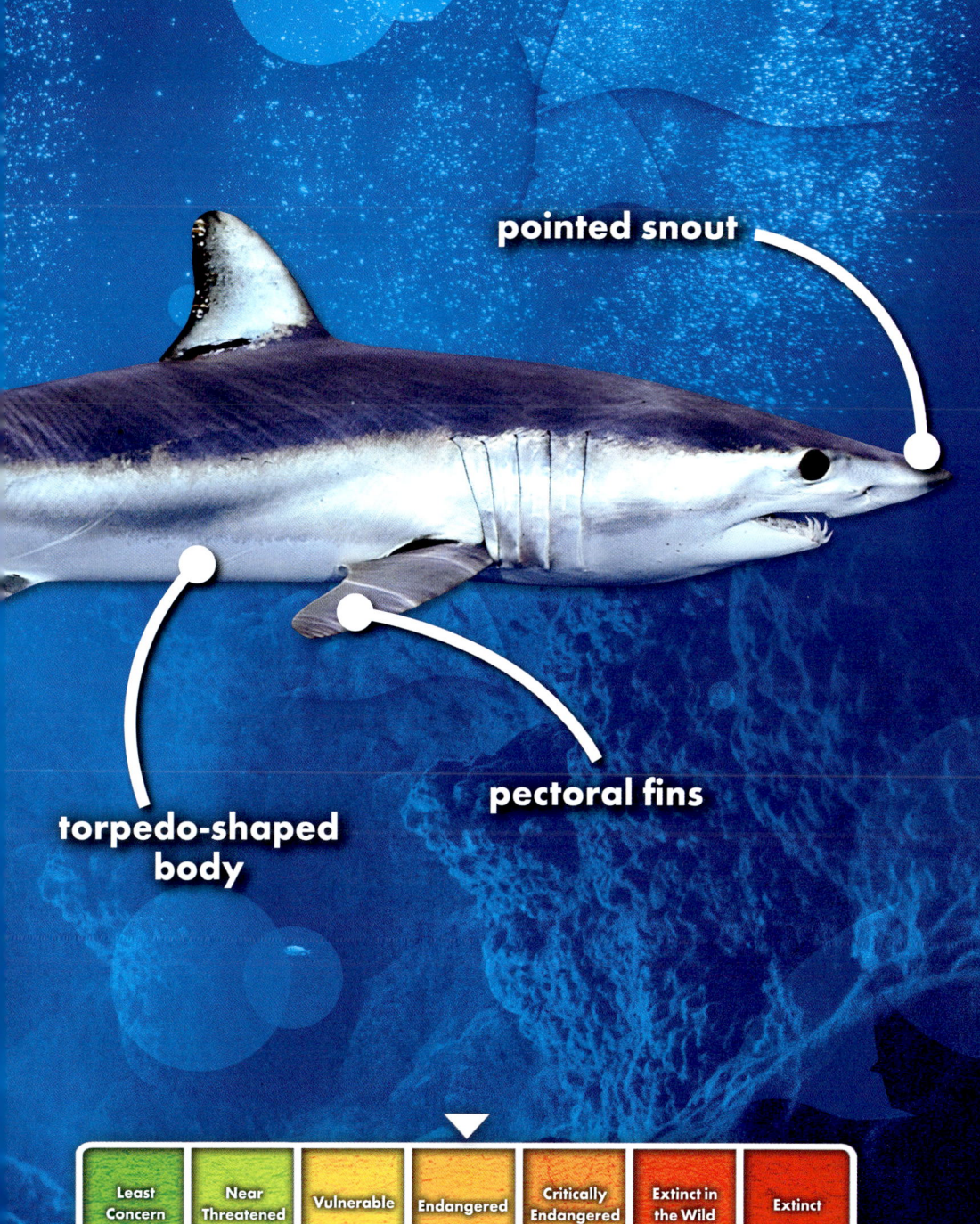

Glossary

apex predators—animals at the top of the food chain that are not preyed upon by other animals

carnivores—animals that only eat meat

endangered—animals or plants that are in danger of dying out

energy—the power to move and do things

gills—parts that help sharks breathe underwater

intelligence—the ability to learn and understand

mackerel sharks—sharks that have two dorsal fins, five gill openings, and a mouth that usually reaches past their eyes; makos and great whites are mackerel sharks.

orcas—killer whales

pectoral fins—a pair of fins on the side of a shark that control a shark's movement

prey—animals that are hunted by other animals for food

snouts—the noses of some animals

species—a group of animals or plants that are similar and can reproduce

torpedo—a tube-shaped weapon fired underwater

To Learn More

AT THE LIBRARY

Adamson, Thomas K. *Great White Sharks*. Minneapolis, Minn.: Bellwether Media, 2021.

O'Daly, Anne. *Sharks*. Tucson, Ariz.: Brown Bear Books, 2020.

Silverman, Buffy. *Mako Sharks in Action*. Minneapolis, Minn.: Lerner Publications, 2018.

ON THE WEB

FACTSURFER

Factsurfer.com gives you a safe, fun way to find more information.

1. Go to www.factsurfer.com.
2. Enter "mako sharks" into the search box and click 🔍.
3. Select your book cover to see a list of related content.

Index

apex predators, 18
backs, 10
bellies, 10
bite, 16
bodies, 9, 12
breathe, 10
carnivores, 5, 19
colors, 10
deep dive, 20-21
energy, 13
eyes, 6
food, 13, 15
gills, 10, 11
hunt, 14, 15
intelligence, 14
mackerel sharks, 8, 9
mouths, 11
oceans, 5, 19
orcas, 18

overfishing, 7
pectoral fins, 6, 9
prey, 14, 15, 16, 18
range, 5
shape, 9
size, 10
snouts, 9, 11
species, 6, 7
speed, 4, 9, 14
status, 7
swim, 12, 13, 14, 19
tails, 9, 16
teeth, 11, 16
water, 9, 12

The images in this book are reproduced through the courtesy of: wildestanimal, front cover, pp. 4-5, 9, 23; Nature Picture Library/ Alamy, pp. 3, 8, 16 (inset); Martin Prochazkacz, p. 6; Image Source/ Alamy, pp. 6-7; Matt_Potenski, p. 11; Mark Conlin/ Alamy, p. 12; saulty72, pp. 12-13, 18-19; Marko Steffensen/ Alamy, pp. 14-15; FtLaudGirl, p. 15 (fish); PeakMystique, p. 15 (dolphins); Carl Salonen, p. 15 (squids); National Geographic Image Collection/ Alamy, pp. 16-17, 20-21; spatuletail, p. 19.